Solar Eclipse Discoveries

A Guide to Enhance Your Experience on Chasing Celestial Wonders

Calista Perry

All rights reserved. No part of this book may be reproduced, distributed, or transmitted in any form or by any means, including photocopying, recording, or other electronic or mechanical methods, without the prior written permission of the copyright owner except in the case of brief quotations embodied in critical reviews and certain other noncommercial uses permitted by copyright.

Copyright © 2024 by Calista Perry

Table of Content

Introduction.. 5

Chapter 1: Definition of Solar Eclipse.... 9

Chapter 2: Scientific Insights............... 13

Chapter 3: Astrological Perspectives... 22

Chapter 4: Cultural Significance.......... 28

Chapter 5: Environmental Effects....... 34

Chapter 6: Practical Considerations for Viewing.. 40

Chapter 7: Driving Tips During Solar Eclipse... 45

Conclusion.. 50

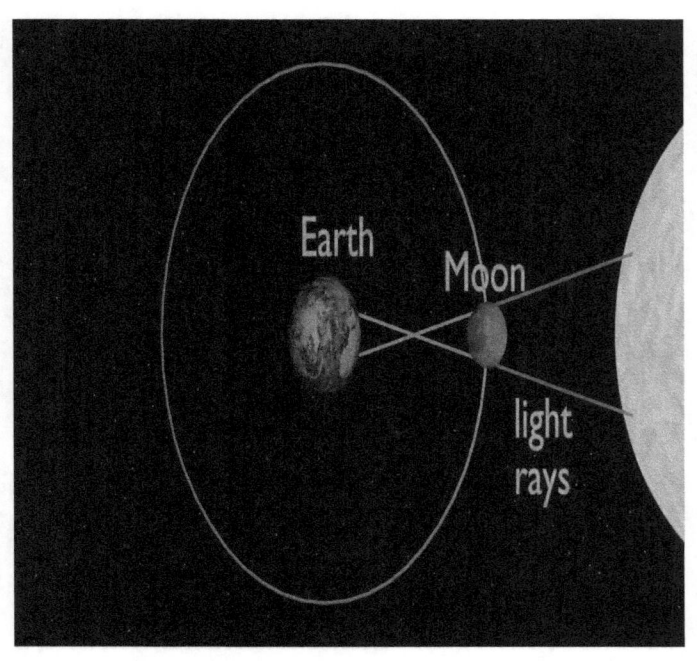

Introduction

Welcome to the exciting world of solar eclipses! In this book, we embark on an exploration of one of nature's most awe-inspiring phenomena, delving into its scientific, astrological, cultural, and environmental dimensions. From ancient myths and legends to modern-day scientific research, solar eclipses have captured the imagination of humanity for millennia, serving as sources of wonder, inspiration, and discovery.

Throughout history, civilizations around the world have observed solar eclipses with a mixture of fear, reverence, and curiosity. From the ancient Greeks and Chinese to indigenous cultures across the globe, people have crafted myths, legends, and rituals to explain and interpret these rare celestial events. In this book, we look at the cultural beliefs and practices surrounding solar eclipses, exploring their significance in different societies and traditions.

From a scientific perspective, solar eclipses offer invaluable opportunities for research and observation. Organizations like NASA conduct detailed studies of solar phenomena during eclipses, using advanced instrumentation and technology to unlock the secrets of the sun's corona and its effects on Earth's atmosphere. By studying solar eclipses, scientists can gain insights into the dynamics of our solar system and deepen our understanding of the universe.

Astrologically, solar eclipses are seen as potent symbols of new beginnings, energy shifts, and personal growth. They are believed to herald transformative changes and offer opportunities for reflection, introspection, and spiritual evolution. In this book, we explore the astrological perspectives on solar eclipses and examine their significance in shaping individual destinies and collective consciousness.

Environmentally, solar eclipses have observable effects on the natural world, from temperature drops and changes in wind patterns to alterations in animal behavior and plant physiology. By studying these effects, scientists can gain insights into the interconnectedness of Earth's ecosystems and the resilience of life on our planet.

In this comprehensive exploration of solar eclipses, we journey through the various facets of these celestial phenomena, from their historical roots to their contemporary significance. Whether viewed through the lens of science, astrology, culture, or the environment, solar eclipses offer profound insights into the beauty, complexity, and interconnectedness of the universe.

Join us on this enlightening journey as we unravel the mysteries of solar eclipses and deepen our appreciation for the wonders of the cosmos. Through exploration, observation, and reflection, we can gain a deeper understanding of our place in the universe and cultivate a sense of wonder and awe for the marvels that surround us.

Chapter 1: Definition of Solar Eclipse

In the vast canvas of the cosmos, few events capture the imagination and curiosity of humanity quite like a solar eclipse. Defined by the intricate dance of celestial bodies—the sun, moon, and Earth—a solar eclipse unfolds as a captivating spectacle, veiling the brilliance of the sun in a moment of celestial drama. In this exploration, we embark on a journey to unravel the mysteries and marvels of solar eclipses, delving into the scientific, cultural, and practical dimensions that shape our understanding and experience of these extraordinary phenomena.

At the heart of our exploration lies the essence of a solar eclipse—the enchanting interplay between the sun, moon, and Earth. A solar eclipse occurs when the moon, in its celestial orbit, aligns precisely between the sun and

Earth, casting its shadow upon our planet. This alignment results in the moon momentarily obscuring the sun's radiant light, transforming day into an ethereal twilight.

The Latin origin of the word "eclipse," derived from "eclipsis," meaning "to fail to appear," aptly captures the essence of this celestial event. As the moon's shadow traverses the Earth's surface, it engenders a profound sense of wonder and awe, prompting cultures throughout history to interpret and imbue solar eclipses with symbolic significance.

Historical Background

Since time immemorial, solar eclipses have captivated the human imagination, inspiring awe, fear, and reverence across diverse cultures and civilizations. Among the earliest documented records of solar eclipses are those of the ancient Greeks, who observed and recorded these celestial events as early as the

8th century BCE. These early astronomers laid the groundwork for subsequent scientific inquiry into the nature and mechanics of solar eclipses.

Throughout history, solar eclipses have been intertwined with cultural beliefs, rituals, and mythologies, shaping the collective consciousness of societies around the world. In ancient China, for instance, solar eclipses were interpreted as celestial omens, signaling cosmic disturbances or divine interventions. The mythological narrative of a dragon devouring the sun reflected the ancient Chinese belief in cosmic forces and the cyclical nature of celestial phenomena.

Similarly, in the annals of Hindu mythology, solar eclipses were attributed to the cosmic struggle between divine entities, symbolizing the eternal conflict between light and darkness. The tale of Rahu, the immortal demon whose severed head caused eclipses by devouring the sun, exemplifies the rich tapestry of cultural narratives woven around these celestial events.

As civilizations evolved and scientific understanding advanced, solar eclipses assumed new significance as opportunities for astronomical observation and exploration. From the pioneering work of astronomers like Sir Arthur Eddington, who confirmed Einstein's theory of general relativity during a solar eclipse in 1919, to the ongoing research conducted by organizations like NASA, solar eclipses continue to yield valuable insights into the nature of the cosmos and our place within it.

In the following sections, we will delve deeper into the scientific, cultural, and practical dimensions of solar eclipses, exploring the myriad ways in which these celestial phenomena shape our understanding of the universe and our place within it.

Chapter 2: Scientific Insights

Solar eclipses serve as celestial laboratories, offering scientists a rare opportunity to unlock the secrets of the sun, probe the mysteries of the universe, and deepen our understanding of Earth's atmosphere. In this exploration of scientific insights, we delve into the significance of solar eclipses for scientific discoveries, NASA's observational projects, research on the sun's corona, and the effects on Earth's atmosphere.

Importance of Solar Eclipse for Scientific Discoveries

Solar eclipses have long been heralded as celestial events of immense scientific importance, providing astronomers and researchers with unique opportunities to study the sun's outer atmosphere, known as the corona. The corona, normally obscured by the sun's intense brightness, becomes visible during a total solar eclipse, allowing scientists to observe and analyze its structure, dynamics, and behavior in unprecedented detail.

One of the key scientific achievements facilitated by solar eclipses is the discovery and study of new elements and phenomena within the sun's corona. From the identification of helium during a solar eclipse in 1868 to groundbreaking research on solar wind, magnetic fields, and solar flares, solar eclipses have played a pivotal role in advancing our understanding of solar physics and astrophysics.

Moreover, solar eclipses offer scientists a unique opportunity to study Earth's atmosphere under rare and unusual conditions. By measuring changes in temperature, humidity, and atmospheric composition during an eclipse, researchers can gain valuable insights into the complex interactions between the sun, Earth, and the atmosphere, shedding light on phenomena such as atmospheric ionization, cloud formation, and atmospheric dynamics.

NASA's Observational Projects

As a leading agency in space exploration and scientific research, NASA has spearheaded numerous observational projects during solar eclipses, harnessing cutting-edge technology and innovative instruments to capture data and imagery of the sun's corona and its effects on Earth's atmosphere.

One such project is the deployment of WB-57 research airplanes equipped with specialized cameras and instruments to observe and photograph the solar corona from high altitudes. These aircraft, capable of operating at 15,000 meters or more above Earth's surface, provide scientists with a unique vantage point to study the intricate structures and dynamics of the corona during a total solar eclipse.

In addition to aerial observations, NASA conducts research using sounding rockets, which are designed to make brief trips into Earth's upper atmosphere to collect data and perform scientific experiments. These rockets, launched at strategic intervals before, during, and after a solar eclipse, enable researchers to study the effects of solar radiation on Earth's ionosphere and atmospheric dynamics.

Furthermore, NASA collaborates with international partners, research institutions, and citizen scientists to coordinate observation campaigns, collect data, and analyze the scientific insights gleaned from solar eclipses. By harnessing the collective expertise and resources of the global scientific community, NASA's observational projects during solar eclipses contribute to our collective understanding of the sun, Earth, and the broader universe.

Research on the Sun's Corona

At the heart of solar eclipse research lies the enigmatic solar corona—a region of tenuous plasma extending millions of kilometers into space, enveloping the sun and permeated by magnetic fields, solar wind, and dynamic phenomena. Despite its significance, the corona remains one of the least understood regions of the sun, posing fundamental questions about its origin, structure, and behavior.

During a total solar eclipse, the moon's alignment with the sun provides a rare opportunity to observe the corona with unprecedented clarity and resolution. Scientists deploy a variety of instruments and techniques to study the corona's temperature, density, and magnetic field, shedding light on its complex dynamics and evolution over time.

One of the primary goals of solar eclipse research is to unravel the mysteries of coronal heating—the process by which the corona reaches temperatures exceeding millions of degrees Celsius, far hotter than the sun's surface itself. By analyzing the dynamics of solar flares, coronal loops, and magnetic reconnection events during eclipses, scientists seek to elucidate the mechanisms responsible for heating the corona and accelerating solar wind.

Moreover, solar eclipses provide valuable opportunities to study coronal mass ejections (CMEs)—explosive releases of solar material into space that can impact Earth's magnetosphere and trigger geomagnetic storms. By monitoring CMEs during eclipses and correlating them with solar activity, scientists can improve space weather forecasting and mitigate the potential impacts of solar storms on telecommunications, navigation systems, and power grids.

Effects on Earth's Atmosphere

In addition to studying the sun's corona, solar eclipses offer scientists a unique opportunity to investigate the effects of solar radiation on Earth's atmosphere, ionosphere, and climate. By measuring changes in temperature, humidity, and atmospheric composition during an eclipse, researchers can gain insights into the complex interactions between solar radiation, atmospheric dynamics, and climate variability.

One of the key effects of a solar eclipse on Earth's atmosphere is the sudden reduction in solar radiation reaching the planet's surface, leading to a temporary cooling of the lower atmosphere. This decrease in solar insolation can cause localized changes in temperature, humidity, and atmospheric pressure, influencing weather patterns and atmospheric circulation on both regional and global scales.

Furthermore, solar eclipses can induce ionospheric disturbances, altering the density, composition, and behavior of charged particles in Earth's upper atmosphere. By monitoring changes in ionospheric parameters such as electron density, ion composition, and plasma irregularities during an eclipse, scientists can gain insights into the dynamics of ionospheric processes and their impact on radio communications, navigation systems, and satellite operations.

Moreover, solar eclipses have been linked to transient changes in atmospheric dynamics, including variations in wind patterns, cloud formation, and atmospheric circulation. By analyzing meteorological data collected before, during, and after an eclipse, researchers can elucidate the mechanisms driving these atmospheric phenomena and their implications for weather forecasting and climate modeling.

In conclusion, solar eclipses serve as celestial laboratories for scientific inquiry, offering unique opportunities to study the sun, Earth, and the broader universe. Through observational projects, research on the sun's corona, and investigations into the effects on Earth's atmosphere, scientists continue to unravel the mysteries of solar eclipses and deepen our understanding of the dynamic interplay between celestial phenomena and terrestrial processes.

Chapter 3: Astrological Perspectives

Solar eclipses have captivated humanity for millennia, not only for their awe-inspiring celestial spectacle but also for their profound astrological significance. In this exploration of astrological perspectives, we delve into the symbolic meanings of solar eclipses as heralds of new beginnings, the energy shifts and spiritual significance they embody, and their impact on personal growth and reflection.

Solar Eclipse as a Symbol of New Beginnings

In astrology, solar eclipses symbolize powerful portals of transformation and renewal, heralding significant shifts in consciousness, destiny, and life path. Occurring at the New Moon phase when the sun and moon align in the same zodiac sign, solar eclipses mark potent moments for setting intentions, initiating projects, and embarking on new journeys.

The darkening of the sun during a solar eclipse represents a temporary veiling of the light, inviting us to delve into the depths of our psyche, confront our shadows, and emerge reborn into a brighter, more illuminated existence. Just as the moon obscures the sun's brilliance, eclipses conceal aspects of our lives that are ripe for transformation, inviting us to release old patterns, beliefs, and behaviors that no longer serve our highest good.

Moreover, the alignment of the sun and moon during a solar eclipse symbolizes the integration of masculine and feminine energies, yin and yang, light and shadow. As the sun's radiant energy is temporarily eclipsed by the moon's shadow, we are invited to honor the sacred balance between solar consciousness (ego, willpower, action) and lunar consciousness (intuition, receptivity, emotion), paving the way for greater harmony and alignment in our lives.

Energy Shifts and Spiritual Significance

Solar eclipses are potent catalysts for energetic shifts and spiritual awakenings, amplifying the vibrational frequencies on Earth and facilitating deeper connections with the divine realms. During an eclipse, the veil between the physical and spiritual dimensions thins, allowing for heightened intuition, psychic insights, and mystical experiences.

Many spiritual traditions view solar eclipses as auspicious times for meditation, prayer, and spiritual practice, as the heightened cosmic energies can facilitate profound inner journeys, soul retrievals, and spiritual activations. By attuning to the frequencies of the eclipse, practitioners can access higher states of consciousness, receive guidance from spirit guides and celestial beings, and accelerate their spiritual evolution.

Furthermore, solar eclipses are believed to activate dormant aspects of our DNA, awakening latent gifts, talents, and abilities that are encoded within our soul's blueprint. As the sun's rays are temporarily obscured, the veils of illusion and limitation are lifted, allowing us to access higher dimensions of consciousness and tap into the infinite reservoir of divine wisdom and creativity.

Impact on Personal Growth and Reflection

In addition to their symbolic and spiritual significance, solar eclipses offer profound opportunities for personal growth, reflection, and self-discovery. By pausing to reflect on the themes and lessons associated with the eclipse's zodiac sign and astrological house, we can gain valuable insights into areas of our lives that are ripe for transformation and evolution.

Solar eclipses often coincide with major life events, transitions, and turning points, signaling the end of one chapter and the beginning of another. Whether it's embarking on a new career path, initiating a creative project, or entering into a transformative relationship, eclipses catalyze growth, expansion, and evolution in all areas of life.

Moreover, solar eclipses invite us to embrace the unknown, surrender to the divine flow of life, and trust in the unfolding of our soul's journey. As the sun's brilliance is temporarily eclipsed, we are reminded of the cyclical nature of existence, the impermanence of form, and the eternal nature of spirit.

In conclusion, solar eclipses are cosmic mirrors reflecting the infinite potentiality of the universe and the boundless possibilities within each of us. As we attune to the symbolic meanings, energy shifts, and spiritual significance of eclipses, we open ourselves to profound experiences of transformation, awakening, and self-realization. May each solar eclipse be a sacred portal of light, guiding us on our journey of soul evolution and spiritual growth.

Chapter 4: Cultural Significance

Solar eclipses have held immense cultural significance throughout history, inspiring awe, reverence, and fascination across diverse civilizations and societies. In this exploration of cultural significance, we delve into the myths and legends surrounding solar eclipses, the cultural practices and traditions associated with these celestial events, and the diverse interpretations they evoke in different societies.

Myths and Legends Surrounding Solar Eclipses

Across cultures and civilizations, solar eclipses have been imbued with mythological narratives and symbolic interpretations, serving as celestial phenomena that elicit wonder and provoke contemplation about the nature of existence and the cosmos.

In ancient China, solar eclipses were often interpreted as the celestial dragon devouring the sun—a mythological motif that symbolized cosmic conflict and renewal. To ward off the dragon and protect the sun, people engaged in rituals involving loud noises, drumming, and ceremonial offerings, reflecting the belief in humanity's role in maintaining cosmic balance.

Similarly, in ancient Mesopotamia, solar eclipses were viewed as omens of divine wrath or impending doom, signaling the displeasure of the gods and the need for appeasement through ritualistic sacrifices and prayers. The Babylonians meticulously recorded solar eclipses on clay tablets, associating them with political upheavals, natural disasters, and dynastic changes.

In Norse mythology, solar eclipses were interpreted as the wolves Skoll and Hati chasing the sun and moon across the sky—a cosmic battle between light and darkness that symbolized the cyclical nature of existence and the eternal struggle between opposing forces.

Cultural Practices and Traditions

Solar eclipses have inspired a myriad of cultural practices and traditions around the world, ranging from ritualistic ceremonies and religious observances to superstitions and taboos aimed at protecting communities from perceived cosmic threats.

In ancient Egypt, solar eclipses were regarded as harbingers of chaos and disorder, prompting the pharaoh to perform rituals to restore cosmic order and appease the gods. Priests and scribes meticulously recorded the timing and duration of eclipses, attributing their occurrence to celestial deities and cosmic cycles.

In Hindu culture, solar eclipses are associated with the mythical demon Rahu, who is believed to swallow the sun as punishment for his transgressions. To prevent Rahu from devouring the sun entirely, Hindus engage in rituals such as bathing in holy rivers, reciting mantras, and making offerings to appease the deity.

In Native American traditions, solar eclipses are viewed as sacred moments of spiritual renewal and communion with the natural world. Tribes across North and South America gather to witness eclipses, performing ceremonial dances, prayers, and songs to honor the cosmic forces at play and seek guidance from the ancestors.

Interpretations in Different Societies

The interpretation of solar eclipses varies widely across different societies, reflecting cultural beliefs, cosmological worldviews, and historical narratives that shape collective consciousness and perception of celestial events.

In modern Western societies, solar eclipses are often viewed through a scientific lens, as awe-inspiring astronomical phenomena that provide opportunities for scientific research, education, and public engagement. Astronomers, educators, and enthusiasts alike gather to observe and study eclipses, using advanced technologies and instruments to capture data and images of these rare celestial occurrences.

In indigenous cultures, solar eclipses are imbued with spiritual significance and cultural meaning, serving as sacred moments of connection with the natural world, ancestral spirits, and cosmic energies. Tribes and communities honor eclipses through ceremonial rituals, storytelling, and communal gatherings that reaffirm their relationship to the land, sky, and cosmos.

In conclusion, solar eclipses are richly layered phenomena that transcend scientific explanations, inviting us to explore the intersection of myth, culture, and cosmology. As we delve into the myths and legends, cultural practices and traditions, and diverse interpretations surrounding solar eclipses, we gain a deeper appreciation for the universal human experience of wonder, awe, and reverence in the face of the mysteries of the cosmos. May each eclipse be a reminder of our interconnectedness with the cosmos and the profound beauty of the universe we inhabit.

Chapter 5: Environmental Effects

Solar eclipses, while captivating celestial events, also exert profound effects on the environment, triggering changes in temperature, influencing animal behavior, and affecting plant life. In this exploration of environmental effects, we delve into the temperature drop and wind pattern changes, observe animal behavior during eclipses, and examine the impact on plant life and photosynthesis.

Temperature Drop and Wind Pattern Changes

One of the most noticeable environmental effects of a solar eclipse is the sudden drop in temperature and alterations in wind patterns experienced in the eclipse's path. As the moon obscures the sun, the reduction in solar radiation leads to a decrease in surface temperatures. The magnitude of this temperature drop depends on various factors, including the duration of totality, atmospheric conditions, and geographical location.

During a total solar eclipse, temperatures can plummet by several degrees Fahrenheit, creating a noticeable chill in the air and altering the ambient climate. The sudden cooling effect is particularly pronounced in regions within the path of totality, where the sun is completely obscured by the moon for a brief period.

In addition to the temperature drop, solar eclipses can also influence wind patterns, causing shifts in atmospheric circulation as the Earth responds to changes in solar heating. The cooling of the Earth's surface during an eclipse can disrupt existing wind patterns, leading to fluctuations in air movement and atmospheric dynamics.

Animal Behavior During Eclipse

Solar eclipses have long fascinated scientists and observers due to their profound impact on animal behavior. From domestic pets to wild animals, many species exhibit distinctive reactions to the sudden darkness and altered light conditions during an eclipse.

Farm animals such as cows, chickens, and horses may display behaviors typically associated with nighttime, such as seeking shelter, roosting, or grazing less actively. Domestic pets, including dogs and cats, may also exhibit signs of confusion or agitation in response to the unusual darkness and changes in ambient light.

In the wild, animals are known to react to solar eclipses in various ways, with some species becoming more vocal or active while others seek refuge and remain hidden. Birds, for example, may alter their singing patterns or exhibit flocking behavior as they navigate the altered light conditions during an eclipse. Similarly, nocturnal animals such as bats and owls may emerge prematurely from their daytime roosts, mistaking the eclipse-induced darkness for nightfall.

Marine life is also affected by solar eclipses, with fish and other aquatic organisms responding to changes in light intensity and temperature. Some species may alter their feeding patterns or exhibit unusual behavior during an eclipse, while others may seek shelter in deeper waters or remain hidden until normal light conditions are restored.

Impact on Plant Life and Photosynthesis

In addition to influencing animal behavior, solar eclipses can have significant effects on plant life and photosynthetic activity. As the sun's rays are partially or completely blocked during an eclipse, the process of photosynthesis—a crucial biological process by which plants convert sunlight into energy—may be disrupted.

During a solar eclipse, the reduction in solar radiation can lead to a temporary slowdown or cessation of photosynthetic activity in plants, resulting in decreased growth rates and metabolic activity. The extent of this impact depends on factors such as the duration of the eclipse, the species of plants involved, and the availability of alternative energy sources.

While some plants may be able to compensate for the temporary reduction in sunlight by increasing photosynthetic efficiency or reallocating resources, others may experience stress or damage as a result of the sudden change in light conditions. In agricultural settings, solar eclipses can pose challenges for crop production and yield, particularly if they coincide with critical stages of plant development such as flowering or fruiting.

In conclusion, solar eclipses exert a range of environmental effects that extend beyond the celestial spectacle itself, influencing temperature, wind patterns, animal behavior, and plant life. By observing and studying these effects, scientists gain valuable insights into the complex interactions between the Earth, sun, and living organisms, highlighting the interconnectedness of the natural world and the profound impact of celestial events on terrestrial ecosystems.

Chapter 6: Practical Considerations for Viewing

As the excitement builds for the upcoming solar eclipse, it's essential to take practical considerations into account to ensure a safe and enjoyable viewing experience. From safety precautions and eye protection to selecting the optimal viewing location and planning ahead for accommodations and travel, careful preparation is key to making the most of this celestial event.

Safety Precautions and Eye Protection

First and foremost, safety should be the top priority when viewing a solar eclipse. Looking directly at the sun without adequate eye protection can cause permanent eye damage or even blindness. Therefore, it's crucial to use proper eye protection, such as solar eclipse glasses or handheld solar viewers, certified to meet international safety standards.

Solar eclipse glasses are specially designed to filter out harmful ultraviolet, visible, and infrared radiation, allowing observers to safely view the sun during an eclipse. When selecting eclipse glasses, be sure to check for the ISO 12312-2 certification, which ensures that the glasses meet the necessary safety requirements. Additionally, inspect the glasses for any signs of damage or wear and discard any that are scratched, torn, or expired.

In addition to eye protection, it's essential to educate yourself and others about the potential risks of viewing a solar eclipse without proper precautions. Avoid using homemade or improvised viewing devices, such as sunglasses or smoked glass, as they do not provide adequate protection against solar radiation and can increase the risk of eye injury.

Choosing the Best Viewing Location

Selecting the right viewing location is crucial for maximizing your experience of the solar eclipse. Ideally, you'll want to be within the path of totality, where the moon completely covers the sun and creates the awe-inspiring phenomenon known as the "totality." However, if traveling to the path of totality is not feasible, you can still observe a partial eclipse from outside this area.

When choosing a viewing location, consider factors such as weather conditions, accessibility, and proximity to amenities. Look for open spaces with unobstructed views of the sky, such as parks, fields, or beaches, where you can set up your viewing equipment and enjoy the eclipse without interference from buildings or trees.

Keep in mind that popular viewing locations may become crowded as the eclipse approaches, so arrive early to secure a good spot and avoid last-minute congestion. If traveling to a remote or unfamiliar area, plan your route in advance and allow extra time for traffic delays or unexpected obstacles.

Planning Ahead for Accommodations and Travel

Given the widespread interest in solar eclipses, accommodations and travel arrangements can fill up quickly, especially in areas near the path of totality. Therefore, it's essential to plan ahead and book your accommodations well in advance to ensure availability and avoid disappointment.

Start by researching potential lodging options, such as hotels, campgrounds, or vacation rentals, in or near the path of totality. Consider factors such as proximity to prime viewing locations, amenities, and pricing when making your decision. Keep in mind that prices may be higher than usual during peak eclipse times, so budget accordingly.

In addition to accommodations, consider your transportation needs and plan your travel itinerary accordingly. Whether driving, flying, or taking public transportation, allow extra time for potential delays or traffic congestion, especially on the day of the eclipse. Consider carpooling or using alternative modes of transportation to reduce traffic congestion and minimize environmental impact.

By taking these practical considerations into account, you can ensure a safe, enjoyable, and memorable viewing experience for the upcoming solar eclipse. Remember to prioritize safety, choose the best viewing location, and plan ahead for accommodations and travel to make the most of this rare celestial event.

Chapter 7: Driving Tips During Solar Eclipse

As excitement builds for the upcoming solar eclipse, it's essential to prioritize safety, especially for those who will be driving during this celestial event. With the sudden changes in lighting conditions and potential distractions, drivers need to take extra precautions to ensure a safe journey. In this section, we'll explore the importance of headlights, awareness of surroundings and road safety, and the risks associated with pulling off on the shoulder during a solar eclipse.

Importance of Headlights

One of the most critical driving tips during a solar eclipse is to ensure that your headlights are turned on, even during daylight hours. As the moon moves between the sun and the earth, the sudden darkness can create visibility challenges for drivers, similar to driving at dusk or dawn. Turning on your headlights not only helps you see better but also makes your vehicle more visible to other drivers on the road.

In addition to headlights, consider using your hazard lights or blinkers to alert other drivers to your presence, especially if you need to slow down or pull off to the side of the road. By increasing your visibility, you can reduce the risk of accidents and ensure a safer driving experience for yourself and others.

Awareness of Surroundings and Road Safety

During a solar eclipse, it's essential to remain vigilant and aware of your surroundings at all times. Keep an eye out for any vehicles that may be pulled off on the side of the road, as well as pedestrians or wildlife that may be affected by the sudden darkness. Use extra caution when changing lanes or passing other vehicles, and be prepared to adjust your driving behavior as needed to accommodate changing road conditions.

It's also crucial to obey all traffic laws and regulations, including posted speed limits and lane restrictions. Avoid distractions such as texting or using your phone while driving, as these activities can significantly impair your ability to react quickly in an emergency. Stay focused on the road ahead and maintain a safe following distance from the vehicle in front of you to allow for adequate stopping time.

Risks of Pulling Off on the Shoulder

While it may be tempting to pull off on the shoulder to view the solar eclipse, doing so can pose significant risks to your safety and the safety of others. In many areas, stopping or parking on the shoulder of a highway or interstate is illegal and can result in fines or penalties. Even in areas where it is permitted, pulling off on the shoulder can increase the risk of accidents, especially if other drivers are not expecting to encounter parked vehicles on the roadway.

Furthermore, pulling off on the shoulder during a solar eclipse can create traffic congestion and impede the flow of vehicles, making it more difficult for emergency responders to reach their destinations. In the event of an accident or medical emergency, every second counts, and delays caused by unnecessary traffic congestion can have serious consequences.

Instead of pulling off on the shoulder, consider finding a safe and designated viewing area where you can park your vehicle and watch the eclipse without endangering yourself or others. Look for public parks, rest areas, or other designated viewing locations where you can safely enjoy the spectacle without disrupting traffic flow or violating any laws.

By following these driving tips during a solar eclipse, you can help ensure a safer and more enjoyable experience for yourself and others on the road. Remember to use your headlights, stay aware of your surroundings, and avoid pulling off on the shoulder to minimize the risk of accidents and ensure a smooth journey during this rare celestial event.

Conclusion

As we conclude our exploration of the multifaceted nature of solar eclipses, it's evident that these celestial events hold significance across scientific, astrological, cultural, and environmental realms. Solar eclipses captivate our imagination, inspire awe, and provide opportunities for reflection and discovery. In this final section, we reflect on the diverse insights gained from our journey through the various aspects of solar eclipses and look ahead to future observations and discoveries.

Throughout history, solar eclipses have fascinated humanity, serving as sources of wonder, inspiration, and scientific inquiry. From ancient civilizations to modern-day astronomers, people have sought to understand and interpret the mysteries of these rare celestial phenomena. Our exploration has revealed the diverse perspectives and interpretations of solar eclipses, ranging from scientific insights into the structure of the sun to astrological beliefs in their spiritual significance.

From a scientific standpoint, solar eclipses offer invaluable opportunities for research and discovery. NASA and other scientific organizations have conducted observational projects to study the sun's corona, analyze its effects on Earth's atmosphere, and advance our understanding of solar phenomena. These efforts have led to groundbreaking discoveries and have shed light on the dynamic interactions between the sun, moon, and Earth.

Astrologically, solar eclipses are seen as potent symbols of new beginnings, energy shifts, and personal growth. They are believed to herald transformative changes and offer opportunities for reflection, introspection, and spiritual evolution. Across cultures and traditions, solar eclipses have been interpreted as auspicious moments for setting intentions, releasing the past, and embracing the future with optimism and hope.

Culturally, solar eclipses have inspired myths, legends, and rituals that reflect humanity's awe and reverence for the cosmos. From ancient China's belief in a dragon devouring the sun to India's myth of the demon Rahu swallowing the sun, these cultural narratives provide rich tapestries of meaning and symbolism. Solar eclipses have also been occasions for communal gatherings, celebrations, and ceremonies, uniting people in shared experiences of wonder and awe.

Environmentally, solar eclipses have observable effects on the natural world, from temperature drops and changes in wind patterns to alterations in animal behavior and plant physiology. These phenomena highlight the interconnectedness of Earth's ecosystems and the intricate ways in which living organisms respond to changes in their environment. By studying these effects, scientists can gain insights into the resilience and adaptability of life on our planet.

As we look ahead to future observations and discoveries, it's clear that solar eclipses will continue to captivate and inspire humanity for generations to come. Advances in technology and scientific instrumentation will enable researchers to study solar phenomena with unprecedented detail and precision, unlocking new insights into the dynamics of our solar system.

From upcoming eclipses visible from different parts of the world to ongoing research initiatives aimed at unraveling the mysteries of the sun, there are countless opportunities for exploration and discovery on the horizon. By harnessing the collective curiosity and ingenuity of scientists, astronomers, and enthusiasts around the globe, we can deepen our understanding of solar eclipses and their broader significance for humanity and the natural world.

Solar eclipses serve as reminders of the beauty, complexity, and interconnectedness of the universe. Whether viewed through the lens of science, astrology, culture, or the environment, these celestial events offer profound insights into the workings of the cosmos and our place within it. By continuing to explore, observe, and appreciate solar eclipses, we can enrich our understanding of the universe and cultivate a deeper sense of wonder and awe for the mysteries that lie beyond.

www.ingramcontent.com/pod-product-compliance
Lightning Source LLC
Chambersburg PA
CBHW070418230526
45471CB00006B/2876